BethBirdBooks

120 Pages College Rule
Orange and Maroon design

www.ingramcontent.com/pod-product-compliance
Lightning Source LLC
Chambersburg PA
CBHW081003170526
45158CB00010B/2887
* 9 7 8 1 0 8 3 0 6 8 9 1 0 *